■ SCHOLASTIC

StoryTime STEM Folk & Fairy Tales

Immacula A. Rhodes

New York • Toronto • London • Auckland • Sydney
Mexico City • New Delhi • Hong Kong • Buenos Aires

Editor: Maria L. Chang
Cover design: Tannaz Fassihi
Cover illustrator: Jason Dove
Interior design: Michelle H. Kim
Interior illustrations: John Lund

Scholastic Inc., 557 Broadway, New York, NY 10012
ISBN: 978-1-338-31697-1
Copyright © 2019 by Scholastic Inc.
All rights reserved.
Printed in the U.S.A.
First printing, January 2019.

1 2 3 4 5 6 7 8 9 10   131   25 24 23 22 21 20 19

Table of Contents

Learning Poster

Introduction

Fee-fi-fo-fum, where can we find some STEM learning fun? With *StoryTime STEM: Folk & Fairy Tales*, of course! The ten units in this resource feature favorite stories accompanied by fun, meaningful activities designed to help young children build important science, technology, engineering, and math skills. Challenges in each unit motivate children to solve problems, generate design ideas, carry out investigations, test solutions, and observe and analyze results. In addition, children build literacy skills as they read the stories.

After introducing each tale to the class, use the suggestions and questions to jumpstart children's scientific thinking. Divide the class into small groups of two or three children then challenge them to do the activities. Each activity comes with a reproducible planning or recording sheet in which children can plan their experiments and record their observations and results. The activities support the Next Generation Science Standards as well as core reading and math standards for Grades K–2. (See Connections to Standards, page 8.)

What's Inside
Each unit includes the following:

- **Illustrated folk or fairy tale:** This is the focal point of the unit and serves as the theme for the STEM explorations.

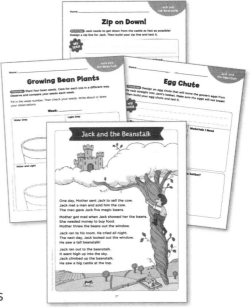

- **Introduction:** Use the ideas and questions in this section to get children thinking like scientists and to encourage discussion.

- **3 hands-on activities:** Each activity poses a story-related challenge for children to complete. Each one includes suggestions for introducing the activity, step-by-step directions, and a list of materials.

- **Planning and recording sheets:** Reproducible sheets help children plan their designs and record observations, data, results, and improvements.

Teaching Tips

Here are some suggestions for getting the most out of your STEM lessons.

- Prior to introducing the stories and activities, gather the materials you'll need to do any suggested demonstrations.

- Enlarge and display the story pages on the board. Track the words as you read the story aloud. Then reread the story, inviting children to read along with you.

- Consider building and displaying a model of your own design to help spark children's creativity. (You'll find illustrations of possible designs for many of the challenges.) Emphasize that your model is only one example of a design that might work to solve the given challenge. Children should try to come up with their own designs and solutions.

- Feel free to add to or substitute materials on the list. Most of the suggested materials are recyclables, common classroom supplies, or are easy to acquire and inexpensive.

- Display the materials and discuss how some might be used in children's designs (e.g., a laundry-bottle lid for a chair leg or a ribbon spool for a vehicle wheel).

- Write or display the activity directions on the board. Go over each step to make sure children understand their task.

- Encourage children to collaborate with their group members to brainstorm ideas about what materials they might use and how they might build their designs to meet the different challenges.

- Help children, as needed, to fill out their planning and recording sheets. Younger children might dictate their design plans for you to jot down, or they might draw pictures for you to label.

- Remind children to use math skills (such as counting, adding, measuring, and estimating) as they plan and build their designs.

- Monitor groups as they work. Ask questions to help prompt critical thinking and problem solving. Encourage children to think out loud about their ideas and solutions.

- After each activity, invite children to discuss their experiences and results. Also, encourage them to share ways in which they might improve their designs to get the desired, or better, results.

- Have a digital camera available to photograph children's designs, experiments, and results. You can then print the images and attach them to children's planning and recording sheets.

- Encourage children to keep a journal of their STEM explorations and results. Provide them with folders in which to store their completed planning and recording sheets, then have them glue a copy of the journal cover (p. 80) to the front. Then invite them to color in the cover.

- Repeat some of the challenges throughout the school year to give children additional opportunities to expand and refine their STEM skills.

- When using water or other liquids, cover the work area with a plastic tablecloth or plastic sheeting. Have paper or cloth towels on hand for cleaning up spills.

- Be aware of any allergies or sensitivities that children might have to certain foods or materials (such as eggs or apples).

Connections to Standards

The activities in this book align with the following standards identified in the Next Generation Science Standards for Grades K–2. For more information, visit www.nextgenscience.org.

PS1.A	Structure and properties of matter	LS2.A	Interdependent relationships in ecosystems
PS2.A	Forces and motion	ESS2.A	Earth materials and systems
PS2.B	Types of interactions	ESS2.E	Biogeology
PS3.C	Relationship between energy and forces	ESS3.A	Natural resources
LS1.A	Structure and function	ESS3.C	Human impacts on Earth systems
LS1.C	Organization for matter and energy flow in organisms	ETS1	Engineering design

The activities also support the College and Career Readiness Anchor Standards for Reading and Mathematics for students in Grades K–2. For more information, visit www.corestandards.org.

Reading

- Demonstrate understanding of the organization and basic features of print.
- Follow words from left to right, top to bottom, and page by page.
- Recognize the distinguishing features of a sentence (e.g., first word, capitalization, ending punctuation).
- Know and apply grade-level phonics and word analysis skills in decoding words.
- Read common high-frequency words by sight.
- Recognize and read grade-appropriate irregularly spelled words.
- Read emergent-reader texts with purpose and understanding.
- Read grade-level text orally with accuracy, appropriate rate, and expression on successive readings.
- Use context to confirm or self-correct word recognition and understanding, rereading as necessary.
- Ask and answer questions about key details in a text.
- Retell familiar stories, including key details.
- Identify characters, settings, and major events in a story.
- Describe characters, settings, and major events in a story, using key details.
- Describe how characters in a story respond to major events and challenges.
- Describe the relationship between illustrations and the story in which they appear.
- Use illustrations and details in a story to describe its characters, settings, or events.
- Compare and contrast the adventures and experiences of characters in familiar stories.
- Actively engage in group reading activities with purpose and understanding.

Mathematics

- Know number names and the count sequence.
- Count to tell the number of objects.
- Compare numbers.
- Understand addition as putting together and adding to, and understand subtraction as taking apart and taking from.
- Represent and solve problems involving addition and subtraction.
- Understand and apply properties of operations and the relationship between addition and subtraction.
- Describe and compare measurable attributes.
- Measure lengths indirectly and by iterating length units.
- Represent and interpret data.
- Measure and estimate lengths in standard units.
- Relate addition and subtraction to length.
- Work with time and money.
- Identify and describe shapes.
- Analyze, compare, create, and compose shapes.
- Reason with shapes and their attributes.

Goldilocks and the Three Bears

Goldilocks came to a house in the woods.
No one was home, so she went in.

Goldilocks saw three bowls of porridge.
She took a bite from each bowl.
The large bowl was too hot.
The medium bowl was too cold.
The small bowl was just right,
so Goldilocks ate it all.

Next, Goldilocks saw three chairs.
She sat in each chair.
The large chair was too high.
The medium chair was too low.
The small chair was just right,
but Goldilocks broke it.

Then Goldilocks saw three beds.
She got into each bed.
The large bed was too hard.
The medium bed was too soft.
But the small bed was just right,
so Goldilocks fell asleep.

After a while, three bears came into the house.
There was a large bear, a medium bear, and a small bear.

The bears looked at the three bowls.
"Someone's been eating my porridge," said all three bears.
Then the small bear said, "And they ate all of mine!"

The bears looked at the three chairs.
"Someone's been sitting in my chair," said all three bears.
Then the small bear said, "And they broke mine!"

The bears looked at the three beds.
"Someone's been sleeping in my bed," said all three bears.
Then the small bear said, "And she is still in mine!"

Just then, Goldilocks woke up.
When she saw the three bears, she jumped out of the bed.
Then she ran out of the house as fast as she could.

The three bears watched Goldilocks run away.
And that was the last time they ever saw her.

StoryTime STEM: Folk & Fairy Tales © Scholastic Inc.

Goldilocks and the Three Bears

INTRODUCTION

Read aloud the fairy tale. Talk about what happens in the story.

To jumpstart scientific thinking, ask children: *How are Goldilocks and the three bears different?* Guide them to notice that their sizes are different—even the bears themselves are different sizes. Note that what might fit just right for the large bear (for example, a chair or bed) may not fit right for the small bear or Goldilocks. Encourage children to think of other ways the bears and Goldilocks may be different from one another and ways they may be similar.

Continue with questions, such as:

- ⚙ What might have caused the small bear's chair to break?
- ⚙ Why do you think Goldilocks didn't feel comfortable in all the chairs and beds?
- ⚙ Why might the bowls of porridge be different temperatures?
- ⚙ What might the bears have done to keep Goldilocks from getting in so easily?

ACTIVITY 1
Just-Right Chair

Challenge: Build a just-right chair for Goldilocks to sit in at the bears' house.

Have children choose a doll or stuffed animal to be their Goldilocks. Tell them they will build a chair that is "just right" for Goldilocks. Explain that one way to make sure something will fit a person just right is to take measurements first. Have children write their ideas and draw their plans on their planning sheet. Then have them do the activity. Afterwards, have them complete their sheet. Discuss which ideas worked, which didn't, and how they can improve their designs.

DIRECTIONS

1 Gather the materials you need.

2 Pick a doll or stuffed animal. This will be your Goldilocks.

3 Build a chair for your Goldilocks. Use your plan from your planning sheet. (Use a ruler or other object for measuring, as needed.)

4 Sit your Goldilocks in the chair.

5 Is the chair just right for Goldilocks? If not, how can you change it to make it fit better?

MATERIALS

- Just-Right Chair planning sheet (p. 13)
- dolls and stuffed animals
- **standard and nonstandard measuring tools:** ruler, measuring tape, yarn, paper strips, craft sticks, crayons
- empty food boxes, tissue boxes, juice cartons, plastic jugs, soda bottles, plastic containers, cardboard tubes, laundry-detergent bottle caps, bubble wrap, packing foam sheets, felt, cotton batting, fabric scraps
- tape, glue, scissors

StoryTime STEM: Folk & Fairy Tales © Scholastic Inc.

ACTIVITY 2
Porridge Cozy

Challenge: Create a cozy that will keep warm porridge warm.

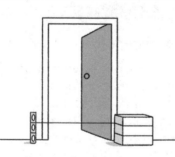

Explain to children that a "cozy" is something that keeps hot things from getting cold. Tell them they will make a cozy to keep a container of warm porridge (water) warm. Ask: *What helps keep you warm?* Have children write their ideas and draw their plans on their planning sheet. Then have them do the activity. Afterwards, have them complete their sheet. Discuss which ideas worked, which didn't, and how they can improve their designs.

DIRECTIONS

1 Gather the materials you need.

2 Make your porridge cozy. Use your plan from your planning sheet.

3 Pour warm water into two cups. Make sure both cups have the same amount of water.

4 Put one cup inside your porridge cozy. Leave the other cup next to it.

5 Wait 10 minutes. Then dip your finger into both cups. Did your porridge cozy keep the water warm?

MATERIALS

- Porridge Cozy planning sheet (p. 14)
- warm water
- 2 large plastic cups (for each child or small group)
- cotton balls, felt, scraps of cloth, wool scarves, clean socks, mittens, paper strips, yarn, bubble wrap
- tape, rubber bands

ACTIVITY 3
Burglar Alarm

Challenge: Design and build a burglar alarm that will make noise when someone enters a room.

Tell children they will build an alarm that will let the bears know when someone enters a room. Ask: *What things make a lot of noise?* Have children write their ideas and draw their plans on their planning sheet. Then have them do the activity. Afterwards, have them complete their sheet. Discuss which ideas worked, which didn't, and how they can improve their designs.

DIRECTIONS

1 Gather the materials you need.

2 Build your alarm. Use your plan from your planning sheet.

3 Set up your alarm by the doorway. Have someone walk in.

4 Did your alarm work? If not, how can you change it to make it work?

MATERIALS

- Burglar Alarm planning sheet (p. 15)
- small bells, empty soda cans, empty plastic bottles, metal jar lids, wooden and plastic blocks, marbles, balloons, craft sticks, paper and plastic cups, metal utensils
- tape, string, rubber bands, glue, scissors

StoryTime STEM: Folk & Fairy Tales © Scholastic Inc.

Name: _____

Just-Right Chair

Challenge Where will Goldilocks sit when she visits the three bears? Design and build a chair that fits Goldilocks. Then test the chair to make sure it fits her just right.

MY IDEA _____
_____

My Plan (Draw here.)

Materials I Need

Did my idea work?

yes

no

How can I make my design better?

Name: _____

Porridge Cozy

Challenge Goldilocks likes her porridge warm. How can she keep it from getting cold? Create a porridge cozy and test it.

MY IDEA _____
_____

My Plan (Draw here.) | **Materials I Need**

Did my idea work?

:-) yes

:-(no

How can I make my design better?

Name: _____

Burglar Alarm

Challenge The bears would like to know when someone enters their house. Design and build an alarm that will make noise when someone comes in through the door.

MY IDEA

_____
_____

My Plan (Draw here.)	**Materials I Need**

Did my idea work?

yes

no

How can I make my design better?

Little Red Riding Hood

One day, Little Red Riding Hood went to see Grandma. Along the way, she met a wolf. "Where are you going, little girl?" asked the wolf.

"I am going to see Grandma," said Little Red. Then off the wolf ran. He wanted to get to Grandma's house first.

When the wolf got there, he locked Grandma in a closet. He put on her gown and glasses. Then the wolf hopped into her bed.

Before long, Little Red walked in. She looked at Grandma. "Grandma," she said, "What big ears you have! And what big eyes and big teeth, too!" Little Red did not know she was talking to the wolf.

StoryTime STEM: Folk & Fairy Tales © Scholastic Inc.

"My dear, I have big ears to hear you better," said the sly wolf. "I have big eyes to see you better. And I have big teeth to eat you up!"

The wolf jumped out of bed. He tried to grab Little Red.

Just then, a man came in. He chased the wolf away. The man told the wolf never to come back again.

The man helped Grandma out of the closet. Little Red and Grandma hugged each other. Then they thanked the man for his help. From that day on, Little Red never spoke to a wolf again.

Little Red Riding Hood

INTRODUCTION

Read aloud the fairy tale. Talk about what happens in the story.

To jumpstart scientific thinking, remind children that Little Red Riding Hood was on her way to see Grandma. Ask children to share their ideas about how Little Red might always be sure to take the quickest path to Grandma's house. Afterwards, explain that Little Red might use a type of "code" to mark the path with special symbols, such as arrows that point out each direction she should take along the way. Invite children to tell about any experiences they might have had with using a code to do an activity or solve a problem.

Continue with questions, such as:

- ✿ What are some things Grandma might do to keep the wolf away from her house?
- ✿ What features might a fence have to prevent the wolf from entering or climbing over it?
- ✿ Can big ears really help you hear better?

ACTIVITY 1
Getting to Grandma's

Challenge: Create a path to help Little Red Riding Hood get to Grandma's house.

Before class, tape a 25-square grid, with 8- to 12-inch squares, on the floor. Place a doll (Little Red Riding Hood) at one corner and a box (Grandma's house) at the opposite corner. Put four or five "trees" (use wooden blocks) in random squares on the grid. Tell children they will create a path from Little Red to Grandma's house. They will use arrows to map out the path, going around the trees as needed. Create your own arrow path and demonstrate how to "walk" Little Red along the path. Then have children work in small groups and complete their recording sheet.

DIRECTIONS

 1 Make a path from Little Red to the house. Use arrows to point the way.

2 Check that your arrows point in the correct direction: → ← ↑ ↓ .

3 "Walk" Little Red along the path. Follow the arrows.

4 Does the path take Little Red to Grandma's house? Correct any arrows, if needed.

5 Now fill in your recording sheet. Draw the path you made.

MATERIALS

- Getting to Grandma's recording sheet (p. 20)
- 25-square grid (with 8- to 12-inch squares) taped to floor
- 15 8-inch tagboard squares, each labeled with an arrow
- wooden blocks or other objects to represent trees
- dolls or stuffed animals
- box (for Grandma's house)
- crayons

ACTIVITY 2
A Wolf-Proof Fence

Challenge: Build a fence that will keep the wolf away from Grandma's house.

Tell children they will build a fence that will keep the wolf away from Grandma's house. Encourage them to consider special features, such as high walls, a spiky top, or a gate that locks. Have children write their ideas and draw their plans on their planning sheet. Then have them do the activity. Afterwards, have them complete their sheet. Discuss which ideas worked, which didn't, and how they can improve their designs.

DIRECTIONS

1 Gather the materials you need.

2 Build your fence. Use your plan from your planning sheet.

3 Now, choose a stuffed animal to represent the wolf.

4 Move the "wolf" around as if it is trying to get inside the fence.

5 Is the fence wolf-proof—will it keep the wolf out? If not, how can you change it to make it better?

MATERIALS

- A Wolf-Proof Fence planning sheet (p. 21)
- cardboard tubes, tall chip canisters, pieces of cardboard, large plastic cups, wooden and plastic blocks, boxes, craft sticks, spring-type clothespins, pipe cleaners
- tape, glue, scissors
- stuffed animals

ACTIVITY 3
My, What Big Ears!

Challenge: Design ears that will help you hear better.

Ask children: *Can big ears really help you hear better?* Tell them they will design ears that can help them hear better. Have children write their ideas and draw their plans on their planning sheet. Then have them work with a partner to test their ears and complete their sheet. Discuss which ideas worked, which didn't, and how they can improve their designs.

DIRECTIONS

1 Gather the materials you need.

2 Build your ears. Use your plan from your planning sheet.

3 Stand in a quiet area with a partner. Take 10 steps away from your partner.

4 Have your partner crinkle a piece of paper. Can you hear it?

5 Now, put on your ears. Repeat Step 4.

6 Did your ears work? If not, how can you change them to make them better?

MATERIALS

- My, What Big Ears! planning sheet (p. 22)
- cardboard tubes, paper or plastic cups of different sizes, cardboard, paper
- tape, scissors
- a partner
- pieces of scrap paper

Name: _____

Getting to Grandma's

Challenge Help Little Red Riding Hood get to Grandma's house.
Use arrows to mark the path.

Draw four or five trees on the grid. Then draw an arrow path from Little Red
to Grandma's house. Check that each arrow points in the correct direction:
→ ← ↑ ↓. Color your path to check that it works.

Extra Challenge What's the shortest way to the house?
Draw the path. How many arrows did you use?

StoryTime STEM: Folk & Fairy Tales © Scholastic Inc.

Name: _____

A Wolf-Proof Fence

Challenge Keep the wolf away from Grandma's house. Design a wolf-proof fence. Then build and test it.

MY IDEA

_____
_____

My Plan (Draw here.)	**Materials I Need**

Did my idea work?

yes

no

How can I make my design better?

Name: _____

My, What Big Ears!

Challenge Can big ears make you hear better? Design ears that will help you hear things far away. Then test them with a friend.

MY IDEA _____
_____

My Plan (Draw here.)

Materials I Need

Did my idea work?

yes

no

How can I make my design better?

Rapunzel

Long ago, a witch took a baby girl from her parents.
The witch named the baby Rapunzel.

Rapunzel grew and grew.
Her hair grew and grew, too.
When Rapunzel turned ten,
the witch locked her in a tower.
No one would ever find her there.

The witch took food to the tower every day.
"Rapunzel, Rapunzel, let down your hair!"
called the witch. Rapunzel dropped
her long hair out the window.
Then the witch climbed up.

One day, a prince took a ride on his horse.
He rode by the tower.
The prince hid when he heard the witch.
He watched the witch climb up Rapunzel's hair.

When the witch left, the prince went to the tower.
"Rapunzel, Rapunzel, let down your hair!" called the prince.
Rapunzel dropped her long hair out the window.
Then the prince climbed up.

Rapunzel was scared when she saw the prince.
But she soon learned he was her friend.
So she told him about the witch.
The prince had a rope.
He used it to help Rapunzel out of the tower.

Then the prince and Rapunzel rode away on his horse.
They rode to the prince's castle, far away from the witch.
And that's where Rapunzel lived, happily ever after.

StoryTime STEM: Folk & Fairy Tales © Scholastic Inc.

Rapunzel

INTRODUCTION

Read aloud the fairy tale. Talk about what happens in the story.

To jumpstart scientific thinking, display the art on the first page of the story. Point out that the witch climbs up and down Rapunzel's long hair to get into and out of the tower. Ask children: *Do you think Rapunzel could use her own hair as a way up and down the tower?* Encourage children to explain their responses. Then invite them to brainstorm ways Rapunzel might get out of the tower for good.

Continue with questions, such as:

✿ How might the witch get food to Rapunzel without climbing up the tower herself?

✿ Might Rapunzel already have what she needs to escape from the tower? Explain.

✿ How might the Prince get up the tower without climbing Rapunzel's hair?

✿ How could a ramp be used to get Rapunzel out of the tower?

ACTIVITY 1
Food Elevator

Challenge: Build an elevator that the witch could use to send food up to Rapunzel.

Tell children they will build an elevator that the witch could use to get food up to Rapunzel. Explain that they will need to use a pulley for their elevator. Display the "Simple Machines" learning poster and point out the pulleys. Have children write their ideas and draw their plans on their planning sheet. Then have them do the activity. Afterwards, have them complete their sheet. Discuss which ideas worked, which didn't, and how they can improve their designs.

DIRECTIONS

1 Gather the materials you need.

2 Build your food elevator. Use your plan from your planning sheet.

3 Place a few pieces of plastic food on the elevator tray.

4 Now raise and lower the food elevator.

5 Does your elevator work? If not, how can you change it to make it work?

MATERIALS

• Simple Machines learning poster (p. 79)

• Food Elevator planning sheet (p. 27)

• **elevator tray:** sturdy paper plates, pie tins, cardboard, shallow boxes, shoebox lids, large frozen-dinner containers

• **pulley:** thread and ribbon spools, empty tape rolls, plastic or cardboard tubes, plastic coffee-cup lids, straws, kite string, yarn, thin rope or cord

• hole punch, tape, paper clips, metal clips, spring-type clothespins, pipe cleaners

• plastic food

ACTIVITY 2
Rapunzel's Parachute

Challenge: Make a parachute to help Rapunzel escape from the tower.

Tell children they will make a parachute that Rapunzel could use to escape from the tower. Explain that the parachute should not drop straight to the ground, but should slow down a bit before it lands. Have children write their ideas and draw their plans on their planning sheet. Then have them do the activity. Afterwards, have them complete their sheet. Discuss which ideas worked, which didn't, and how they can improve their designs.

DIRECTIONS

1 Gather the materials you need.

2 Build your parachute. Use your plan from your planning sheet.

3 Attach your parachute to a small toy.

4 Stand on a stool against a wall. (Or sit at the top of a slide.) Hold out the parachute in front of you and drop it.

5 Does your parachute work? If not, how can you change it to make it work?

MATERIALS

- Rapunzel's Parachute planning sheet (p. 28)
- tissue paper, newsprint, paper towels, waxed paper, thin fabric, coffee filters; kite string, dental floss, thin yarn, thin gift ribbon
- hole punch, tape, scissors
- small dolls or toys

ACTIVITY 3
A Super Slide

Challenge: Create a slide that Rapunzel can use to escape safely from the tower.

Tell children they will build a slide that Rapunzel can go down to get out of the tower. Explain that she should be able to slide down safely, without falling off the sides. Have children write their ideas and draw their plans on their planning sheet. Then have them do the activity. Afterwards, have them complete their sheet. Discuss which ideas worked, which didn't, and how they can improve their designs.

DIRECTIONS

1 Gather the materials you need.

2 Build your slide and set it up. Use your plan from your planning sheet. (Prop up your slide on the edge of a bookshelf or a table or the back of a chair.)

3 Make sure your slide won't slip or shift when it is used.

4 Choose a doll or stuffed animal to be Rapunzel. Now, send Rapunzel down the slide.

5 Does Rapunzel slide to the bottom without falling off? If not, how can you change the slide to make it work?

MATERIALS

- A Super Slide planning sheet (p. 29)
- **slide:** cardboard box lid, 12-pack soda cartons, large rectangular foil pan lids, tall juice cartons (cut in half lengthwise, ends removed), large shoeboxes
- **support:** boxes of various heights, wooden blocks, tall containers, stacks of books, stools, chairs
- **ramps:** sturdy cardboard planks, wood planks
- small dolls or stuffed animals

StoryTime STEM: Folk & Fairy Tales © Scholastic Inc.

Name: _____

Food Elevator

Challenge The witch brings Rapunzel food every day. But it's hard to climb up and down the tower. Design an elevator she can use to send food up the tower. Then build and test your elevator.

MY IDEA _____
_____

My Plan (Draw here.)

Materials I Need

Did my idea work?

:) yes

:(no

How can I make my design better?

Name: _____

Rapunzel's Parachute

Challenge Rapunzel wants to escape from the tower. Design a parachute to help Rapunzel jump safely to the ground. Put your parachute together. Then test it.

MY IDEA _____

_____

My Plan (Draw here.)

Materials I Need

Did my idea work?

yes

no

How can I make my design better?

StoryTime STEM: Folk & Fairy Tales © Scholastic Inc.

Name: _____

A Super Slide

Challenge Everyone enjoys going down a slide—even Rapunzel!
Design a slide that she can use to get down from the tower.
Then build and test your slide.

MY IDEA

_____

_____

My Plan (Draw here.)	**Materials I Need**

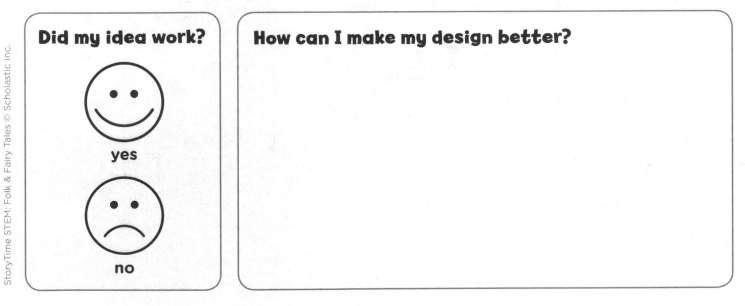

Did my idea work?

yes

no

How can I make my design better?

The Three Little Pigs

Once upon a time, three little pigs set out to build houses.

The first little pig made a straw house.
Along came the big, bad wolf.
The wolf called out, "Little pig, little pig, let me in!"

"Not by the hair on my chinny-chin-chin!" cried the little pig.
So the wolf huffed and puffed.
He blew the straw house down.
The little pig ran away as fast as he could.

The second little pig made a stick house.
Along came the big, bad wolf.
The wolf called out, "Little pig, little pig, let me in."

"Not by the hair on my chinny-chin-chin!" cried the little pig.
So the wolf huffed and puffed.
He blew the stick house down.
The little pig ran away as fast as he could.

The third little pig made a brick house.
Along came the big, bad wolf.
The wolf called out, "Little pig, little pig, let me in."

"Not by the hair on my chinny-chin-chin!" cried the little pig.
So the wolf huffed and puffed.
He blew and blew.

The brick house did not move.
It did not fall down.
Finally, the wolf gave up.
He left and went far, far away.

The other little pigs ran to the brick house.
Then the three little pigs cheered, "Hooray!
The big, bad wolf is gone!"

The Three Little Pigs

INTRODUCTION

Read aloud the fairy tale. Talk about what happens in the story.

To jumpstart scientific thinking, display the art on the second page of the story. Point out that the brick house did not move, no matter how hard the wolf blew. Ask children: *Why do you think the wolf could blow down the straw and stick houses but not the brick house?* Then explain that wind is air that moves. Tell children that when the wolf huffed and puffed and blew, the air coming out of his mouth was very much like wind. Invite children to brainstorm things that work or move because of the wind.

Continue with questions, such as:

- ⚙ Why is it important to build homes and other buildings with sturdy materials?
- ⚙ How can wind be helpful to us? How can it be harmful?
- ⚙ What are some things that you can blow with your own breath?
- ⚙ What other things might the wolf be able to blow down? Why?

ACTIVITY 1
Strong House

Challenge: Design a home that the big, bad wolf can't blow down.

Tell children they will build a house for the three little pigs—one that the big, bad wolf can't blow down. Ask: *How could you keep a house from being blown away?* Have children write their ideas and draw their plans on their planning sheet. Then have them do the activity. Afterwards, have them complete their sheet. Discuss which ideas worked, which didn't, and how they can improve their designs.

DIRECTIONS

1 Gather the materials you need.

2 Build your house for the three little pigs. Use your plan from your planning sheet.

3 Put three toy pigs in the house.

4 Now, pretend to be the big, bad wolf. Blow on the house as hard as you can. (Or use a fan or blow dryer instead.)

5 Blow on the house two or three more times. Does the house stay up? If not, how can you change it so it won't blow away?

MATERIALS

- Strong House planning sheet (p. 34)
- tagboard, plastic and foam cups, cardboard boxes, sturdy paper plates and bowls, cardboard tubes, tall chip canisters, craft sticks, sturdy straws, play dough, tape
- toy pigs or other animals
- fan or blow dryer (optional)

StoryTime STEM: Folk & Fairy Tales © Scholastic Inc.

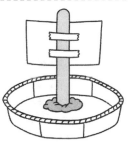

ACTIVITY 2
A Working Windmill

Challenge: Build a windmill that works when the big, bad wolf blows on it.

Tell children they will build a windmill that will put the big, bad wolf's strong blows to good use. Explain that a windmill uses wind power to do work. Have children write their ideas and draw their plans on their planning sheet. Then have them do the activity. Afterwards, have them complete their sheet. Discuss which ideas worked, which didn't, and how they can improve their designs.

DIRECTIONS

1 Gather the materials you need.

2 Build your windmill. Use your plan from your planning sheet.

3 Now, pretend to be the big, bad wolf. Blow on the blades of your windmill. (Or use a fan or blow dryer instead.)

4 Do the blades spin around? If not, how can you change them to make them spin?

MATERIALS

- A Working Windmill planning sheet (p. 35)
- **windmill base:** tall food boxes, narrow shoeboxes, chip canisters, juice cartons, large foam or plastic cups, tall plastic soda bottles, paper-towel tubes
- **blades:** tagboard, cardboard, poster board, foam sheets, Styrofoam trays; paper fasteners, straws, craft sticks, large beads, ribbon spools
- tape, glue, scissors
- fan or blow dryer (optional)

ACTIVITY 3
Wind Works

Challenge: Create something that will work when you blow into it or at it.

Ask children: *What things need air to work or to move?* (Some examples are a horn, whistle, pinwheel, sailboat, or bubble-blower.) Tell children they will create something that moves or works when they blow air into it or at it. Have children write their ideas and draw their plans on their planning sheet. Then have them do the activity. Afterwards, have them complete their sheet. Discuss which ideas worked, which didn't, and how they can improve their designs.

DIRECTIONS

1 Gather the materials you need.

2 Make your wind device. Use your plan from your planning sheet.

3 How is your device supposed to work—by blowing into it or at it? Blow to test it.

4 Does your device work? If not, how can you change it to make it work?

MATERIALS

- Wind Works planning sheet (p. 36)
- foam trays, foil, craft sticks, straws, pipe cleaners, plastic soda bottles, foam and plastic cups, plastic funnels, plastic flip-top lids (from condiment bottles, toothpaste tubes, etc.), plastic tape-roll rings, plastic thread and ribbon spools, metal Mason jar rings, plastic strawberry baskets, cardboard tubes, spring-type clothespins, tagboard
- paper fasteners, tape

Name: _____

Strong House

Challenge Design a house that the big, bad wolf can't blow down. Build your house. Blow on it. Does it stay up?

MY IDEA _____
_____

My Plan (Draw here.)	**Materials I Need**

Did my idea work?

yes

no

How can I make my design better?

Name: _____

A Working Windmill

Challenge Put the big, bad wolf's strong blows to work!
Design and build a windmill that works.

MY IDEA
_____
_____

My Plan (Draw here.)

Materials I Need

Did my idea work?

yes

no

How can I make my design better?

Name: _____

Wind Works

Challenge Create something that works when you blow air into it or at it. Then make and test your idea.

MY IDEA

_____

_____

My Plan (Draw here.) | **Materials I Need**

Did my idea work?

yes

no

How can I make my design better?

Jack and the Beanstalk

One day, Mother sent Jack to sell the cow.
Jack met a man and sold him the cow.
The man gave Jack five magic beans.

Mother got mad when Jack showed her the beans.
She needed money to buy food.
Mother threw the beans out the window.

Jack ran to his room. He cried all night.
The next day, Jack looked out the window.
He saw a tall beanstalk!

Jack ran out to the beanstalk.
It went high up into the sky.
Jack climbed up the beanstalk.
He saw a big castle at the top.

Jack went in the castle. He looked around.
He saw a basket. It was full of golden eggs.
Then Jack saw a goose. It was the goose that laid the eggs!

Just then, a giant came in. He saw Jack and roared,
"Fee-fi-fo-fum! I'm going to eat you, little one!"

Jack was scared, but he moved fast. He grabbed the goose
and the basket. Then down the beanstalk he went.

At the bottom, Jack cut down the beanstalk.
Now the giant had no way to get to Jack!

Jack showed Mother the goose and eggs.
She danced with joy. The goose would make them rich!
Mother and Jack would never go hungry now!

StoryTime STEM: Folk & Fairy Tales © Scholastic Inc.

Jack and the Beanstalk

INTRODUCTION

Read aloud the fairy tale. Talk about what happens in the story.

To jumpstart scientific thinking, display the art on the first page of the story. Point out the bean plant. Tell children that plants grow in the ground. Ask if they know what part of a plant grows underground (roots). Then have children name the plant parts that grow above the ground (stem, leaves, flowers, and/or fruit). Invite children to share what they know about plants and how they grow.

Continue with questions, such as:

- ✿ Do you think it's easier for Jack to climb up or to climb down the bean plant? Why?
- ✿ Other than climbing, how else might Jack get to and from the giant's castle?
- ✿ What might happen if a plant does not get water and light?

ACTIVITY 1
Growing Bean Plants

Challenge: Explore what a seed needs to grow. Care for each of four bean seeds in a different way.

Ask children: *What do you think plants need to grow?* Tell children they will plant four seeds in separate containers to find out. They will care for each seed in a different way. Each week, they will check their seeds and record their observations on their recording sheet. At the end of four weeks, discuss what they observed.

DIRECTIONS

1 Prepare four planters. Fill each planter about ⅔ full with soil. Plant a bean in each planter.

2 Label the planters as follows:

    a. Water only   **b.** Light only   **c.** Water and light   **d.** Nothing

3 Put the "water only" and "nothing" planters in a dark place, such as a closet or under a box.

4 Place the "light only" and "water and light" planters on a sunny windowsill.

5 Add about ¼ cup of water to the "water only" and "water and light" planters. Repeat whenever the soil feels dry.

6 Check each planter once a week. Record your observations on your recording sheet.

7 After four weeks, bring your planters together. Compare them.

MATERIALS

- Growing Bean Plants recording sheet (p. 41), one copy per week for each child
- **planters:** small foam or plastic cups, plastic jars, yogurt cups
- lima beans, soil, water
- markers
- index cards or sticky labels
- measuring cups

ACTIVITY 2
Zip on Down!

Challenge: Make a zip line that Jack can use to get from the giant's castle back down to the ground.

Tell children they will build a zip line for Jack to escape from the giant's castle. Have children write their ideas and draw their plans on their planning sheet. Then have them do the activity. Afterwards, have them complete their sheet. Discuss which ideas worked, which didn't, and how they can improve their designs.

DIRECTIONS

 1 Gather the materials you need.

 2 Build your zip line. Use your plan from your planning sheet.

3 Choose a plastic toy figure or small stuffed animal to represent Jack.

4 Put Jack on your zip line. Send him down.

 5 Does your zip line work? If not, how can you change it to make it work?

MATERIALS

- Zip on Down! planning sheet (p. 42)
- **seat:** small plastic containers, paper and foam bowls and cups, small boxes, strawberry baskets, felt and fabric strips (seatbelt)
- **zip line:** cardboard tubes, ribbon spools, plastic shower-curtain rings, large paper clips, thin rope, ribbon, yarn, tape
- small dolls, plastic figures, and stuffed animals

ACTIVITY 3
Egg Chute

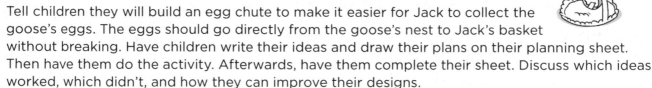

Challenge: Design and build an egg chute that will move the golden eggs safely from the goose's nest to Jack's egg basket.

Tell children they will build an egg chute to make it easier for Jack to collect the goose's eggs. The eggs should go directly from the goose's nest to Jack's basket without breaking. Have children write their ideas and draw their plans on their planning sheet. Then have them do the activity. Afterwards, have them complete their sheet. Discuss which ideas worked, which didn't, and how they can improve their designs.

DIRECTIONS

 1 Gather the materials you need.

 2 Build your egg chute. Use your plan from your planning sheet.

3 Hold a hard-boiled egg in place in the nest at the top of your chute.

 4 Now, let the egg go.

5 Does the egg roll down the chute and into the basket? Does it break? If the egg breaks, how can you change the chute to make it better?

MATERIALS

- Egg Chute planning sheet (p. 43)
- **nest:** paper and foam plates and bowls, shallow boxes; bubble wrap, foam packing peanuts, shredded paper, cotton batting, cotton balls
- **chute:** sheets of sturdy tagboard and cardboard, cereal boxes, shoeboxes, tissue boxes, wooden blocks, stacks of books, sturdy tape
- large baskets or containers to represent baskets
- hard-boiled eggs

Name: _____

Growing Bean Plants

Challenge Plant four bean seeds. Care for each one in a different way. Observe and compare your seeds each week.

Fill in the week number. Then check your seeds. Write about or draw your observations.

Week: _____

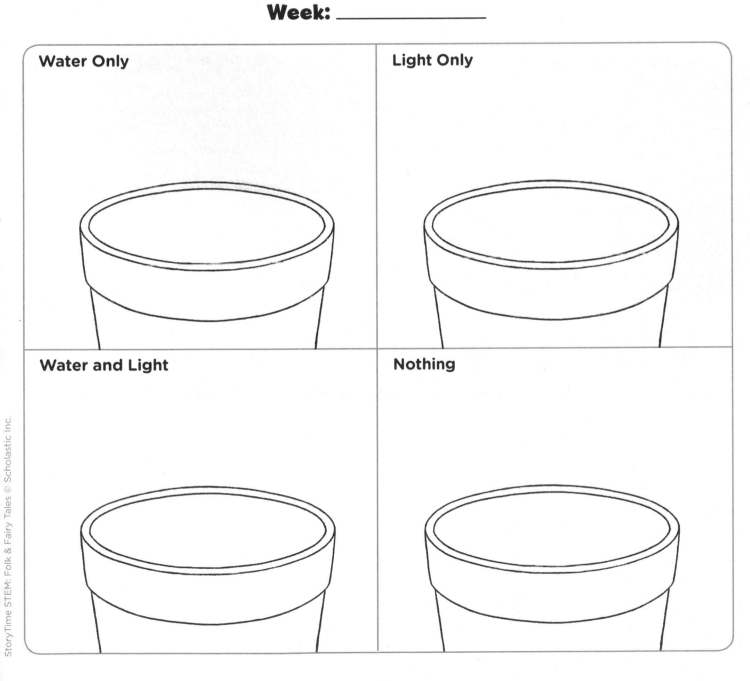

Water Only

Light Only

Water and Light

Nothing

StoryTime STEM: Folk & Fairy Tales © Scholastic Inc.

Name: _____

Zip on Down!

Challenge Jack needs to get down from the castle as fast as possible! Design a zip line for Jack. Then build your zip line and test it.

MY IDEA _____
_____

My Plan (Draw here.)	**Materials I Need**

Did my idea work?

yes

no

How can I make my design better?

StoryTime STEM: Folk & Fairy Tales © Scholastic Inc.

Egg Chute

Challenge Design an egg chute that will move the goose's eggs from its nest straight into Jack's basket. Make sure the eggs will not break! Then build your egg chute and test it.

MY IDEA
_____
_____

My Plan (Draw here.)

Materials I Need

Did my idea work?

yes

no

How can I make my design better?

The Gingerbread Man

One day, a farmer made a gingerbread man.
The gingerbread man came to life.
He jumped off the pan and ran out the door.

The gingerbread man called out, "Run, run as fast as you can.
You can't catch me. I'm the Gingerbread Man!"
The farmer ran after the gingerbread man.

Soon, the gingerbread man ran past a cow.
He ran past a pig. Then he ran past a dog.

The gingerbread man called out, "Run, run as fast as you can.
You can't catch me. I'm the Gingerbread Man!"

So the cow ran after the gingerbread man.
The pig ran after the gingerbread man.
And the dog ran after the gingerbread man.

Before long, the gingerbread man came to
a river. He did not know how he would get across.

StoryTime STEM: Folk & Fairy Tales © Scholastic Inc.

Just then, a fox came along.
The fox said, "Hop onto my back. I will take you across."

So the gingerbread man hopped onto the fox's back.
The fox swam and swam.
The water got deeper and deeper.

The gingerbread man moved to the fox's head.
Then he moved to the fox's nose.

Now the fox could smell the yummy gingerbread man.
The fox licked his lips. All of a sudden . . . *Snap!*
The fox ate the gingerbread man in one big bite.

And the gingerbread man was never seen again.

The Gingerbread Man

INTRODUCTION

Read aloud the fairy tale. Talk about what happens in the story.

To jumpstart scientific thinking, tell children that a gingerbread man is a type of cookie. Invite children to share what they know about making cookies. Talk about what happens when the different ingredients are mixed together and how the dough changes after the cookies have been baked. (You might make gingerbread cookies with children before doing this unit.) Give children gingerbread cookies to taste. Have them share their observations about how their cookie looks, smells, tastes, feels, and sounds as they eat it.

Continue with questions, such as:

⚙ How might the farmer have prevented the Gingerbread Man from running away?

⚙ How might the Gingerbread Man go faster to stay ahead of the animals chasing him?

⚙ How else could the Gingerbread Man have crossed the river?

⚙ Do you think a gingerbread man could float on water? Why or why not? Explain.

ACTIVITY 1
No-Cook Play Dough

Challenge: Explore how solids and liquids are different. Find out what happens when they are mixed together.

Tell children you will be making play dough using some of the same ingredients we use to make cookies. The only difference is you won't be baking or eating it. Instead, children will use the play dough to create their own gingerbread man. (Note: Warn children not to eat the play dough or the gingerbread man.)

Make the play dough: Combine the flour with the salt in a bowl. Invite children to describe the dry ingredients. In a large plastic cup (or measuring cup), combine the cold water and the oil, then add two or three drops each of red and yellow food coloring. Ask children: *How is this mixture different from the flour and salt?* Next, pour the liquid into the bowl and mix together all the ingredients until they are combined. Knead the dough well. Have children describe what happens when you mix all the ingredients together. What does the play dough feel like? Have children record their observations on their recording sheet. Then give each child a piece of play dough and invite children to make their own gingerbread man.

DIRECTIONS

1 Watch your teacher make play dough.

2 Answer the questions on your recording sheet.

3 Make your own gingerbread man. Use the play dough.

MATERIALS

- No-Cook Play Dough recording sheet (p. 48)
- **dry ingredients:** 4 cups plain flour, 2 cups salt
- **wet ingredients**: 2 tablespoons oil, 2 cups cold water, red and yellow food coloring
- mixing bowls, measuring cups and spoons

ACTIVITY 2
A Vehicle for the Gingerbread Man

Challenge: Build a vehicle with at least three wheels for the Gingerbread Man.

Explain to children that riding in a vehicle with wheels is likely faster than running. Tell children they will build a vehicle with at least three wheels for the Gingerbread Man. Have them write their ideas and draw their plans on their planning sheet. Then have them do the activity. Afterwards, have them complete their sheet. Discuss which ideas worked, which didn't, and how they can improve their designs.

DIRECTIONS

1. Gather the materials you need.

2. Build your vehicle. Use your plan from your planning sheet.

3. Pick a stuffed animal to be the Gingerbread Man.

4. Put the Gingerbread Man in the vehicle. Now, roll the vehicle around.

5. Does your vehicle roll? Can it carry the Gingerbread Man? If not, how can you change it to make it work?

MATERIALS

- A Vehicle for the Gingerbread Man planning sheet (p. 49)
- **vehicle body:** assorted gift boxes, tissue boxes, ice cream cartons, plastic food containers, cardboard, foil pans
- **wheels and axles:** ribbon spools, cardboard circles, plastic cup lids; straws, dowels, chopsticks, lollipop sticks
- wide craft sticks, thick cardboard strips, pipe cleaners, paper clips, rubber bands, paper fasteners, hole punch, scissors, tape
- stuffed animals and dolls

ACTIVITY 3
Sink or Float?

Challenge: Explore whether different objects sink or float when they are put in water.

Ask children: *What do you think would happen to the Gingerbread Man if he went into the river? Would he sink or float?* Explain to children that they will choose and test different objects to see if they sink or float. As children do the experiment, have them fill in their recording sheet.

DIRECTIONS

1. Choose an object. Draw a picture of the object on your recording sheet.

2. Make a guess: Will the object sink or float? Check the box for your guess.

3. Place the object in water.

4. Does the object sink or float? Mark the box for your answer.

5. Repeat Steps 1 to 4 with three different objects.

MATERIALS

- Sink or Float? recording sheet (p. 50)
- coins, paper clips, corks, shells, craft sticks, plastic spoons, pencils, plastic blocks, wood blocks, small air-filled balloons, foam and plastic shapes, metal and rubber washers, paintbrushes, feathers, crayons, flat pieces of foil, balled up pieces of foil, small plastic bottles (with and without lids)
- water table or tub of water

StoryTime STEM: Folk & Fairy Tales © Scholastic Inc.

Name: _____

No-Cook Play Dough

Challenge Watch your teacher make play dough.
Then use the play dough to make your own gingerbread man.

Describe the dry ingredients.	**Describe the wet ingredients.**

What happens when your teacher mixes the dry and wet ingredients together?
What does the play dough feel like?

_____

_____

Make your own gingerbread man. Use the play dough.
Draw your gingerbread man below.

Name: _____

A Vehicle for the Gingerbread Man

Challenge Help the Gingerbread Man get away faster! Design a vehicle with at least three wheels. Then build your vehicle and test it.

MY IDEA _____
_____

My Plan (Draw here.)

Materials I Need

Did my idea work?

yes

no

How can I make my design better?

Name: _____

Sink or Float?

Challenge Test different objects to see if they sink or float.

Choose an object. Draw a picture of it. Then put the object in water. Does it sink or float? Mark your answer.

Object (Draw each one below.)	Make a guess: Will it sink or float?	What does it do? Check one.
	☐ Sink ☐ Float	☐ Sink ☐ Float
	☐ Sink ☐ Float	☐ Sink ☐ Float
	☐ Sink ☐ Float	☐ Sink ☐ Float
	☐ Sink ☐ Float	☐ Sink ☐ Float

The Three Billy Goats Gruff

One day, three goats went to cross the river.
They did not know that a troll lived under the bridge.

The smallest goat went first.
Trip-trip, trip-trip, trip-trip.

"Get off my bridge," yelled the troll.
"Or I will come up to eat you!"
The goat ran back to his brothers.

The medium goat went next.
Trip-trap, trip-trap, trip-trap.

"Get off my bridge," yelled the troll.
"Or I will come up to eat you!"
The goat ran back to his brothers.

The largest goat went last.
Trip-trop, trip-trop, trip-trop.

"Get off my bridge," yelled the troll.
"Or I will come up to eat you!"
The goat kept going.
So the troll hopped onto the bridge.

The goat began to run.
He ran right into the troll!
Splash! The troll fell into the water.
The river carried the troll away.

"Hooray!" cheered the three goats.
"The troll is gone.
Now we can all cross the bridge!"

And that's just what the happy goats did.
All three goats crossed the bridge together.
Trip-trip! Trip-trap! Trip-trop!

The Three Billy Goats Gruff

INTRODUCTION

Read aloud the fairy tale. Talk about what happens in the story.

To jumpstart scientific thinking, display the art on the first page of the story. Point out the bridge in the picture. Remind children that the troll hopped up on the bridge to eat the large goat. Then ask children to think about some ways that the bridge might be built to keep the troll off it. Invite them to share and discuss their ideas.

Continue with questions, such as:

✿ How might a covered bridge help protect the goats as they cross it?

✿ What are some other ways the goats might try to cross the river?

✿ How might the troll be removed from the bridge without hurting it?

✿ Why do you think the river was able to carry the troll away?

ACTIVITY 1

Covered Bridge

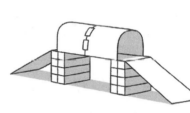

Challenge: Build a covered bridge that will keep the goats safe from the troll.

Tell children they will build a covered bridge so the troll can't jump onto the pathway. Explain that their bridge should be able to support all three goats at the same time. Have children write their ideas and draw their plans on their planning sheet. Then have them do the activity. Afterwards, have them complete their sheet. Discuss which ideas worked, which didn't, and how they can improve their designs.

DIRECTIONS

1 Gather the materials you need.

2 Build your covered bridge. Use your plan from your planning sheet.

3 Draw a small, medium, and large goat on paper. Cut out your goats. Tape each one to a small cube.

4 "Walk" all three goats across your bridge together. (You might place them on a long strip of paper and slide the paper through the covered bridge.)

5 Does your bridge support the weight of all the goats? If not, how can you change it to make it work?

MATERIALS

• Covered Bridge planning sheet (p. 55)

• building blocks, stacks of books, large hardcover books, sturdy cardboard boxes in a variety of shapes and sizes, paper box lids, flat plastic food trays, long foam trays, thin wooden planks, large plastic cups, tall chip canisters, large sheets of construction paper, sheets of foil, foam sheets

• paper, crayons, scissors, one-inch cubes, tape

A Goat Boat

Challenge: Make a boat that will get all three goats across the river safely.

Ask children: *Is there another way the goats can cross the river without using the bridge?* Tell them they will build a boat for the goats. Explain that their boat should be large enough to hold all three goats and stay afloat as it crosses a tub of water. Have children write their ideas and draw their plans on their planning sheet. Then have them do the activity. Afterwards, have them complete their sheet. Discuss which ideas worked, which didn't, and how they can improve their designs.

DIRECTIONS

1 Gather the materials you need.

2 Build your boat. Use your plan from your planning sheet.

3 Place three small plastic goats in the boat. (Or use any other small plastic animals.)

4 Put your boat in water to test it. Does your boat hold all three goats? Does it float long enough to cross the water? If not, how can you change it to make it work?

MATERIALS

- A Goat Boat planning sheet (p. 56)
- plastic bowls, milk cartons, foil pans, plastic food containers, play dough, egg cartons, foam trays, foam blocks, corks
- rubber bands, pipe cleaners, craft sticks, paper (for sails), tape
- small plastic goats or other animals
- water table or tub of water

Troll Trap

Challenge: Design and build a trap to capture the troll.

Tell children they will build a trap so they can capture the troll and move him far away where he can't cause trouble. Explain that their trap should surround the troll when he is captured in it. Have children write their ideas and draw their plans on their planning sheet. Then have them do the activity. Afterwards, have them complete their sheet. Discuss which ideas worked, which didn't, and how they can improve their designs.

DIRECTIONS

1 Gather the materials you need.

2 Build your trap. Use your plan from your planning sheet.

3 Choose a stuffed animal or doll to represent the troll. Make sure the troll is the right size to fit inside your trap.

4 Now, test your trap. Capture the troll in it.

5 Does your trap work? Does it surround the captured troll? If not, how can you change it to make it work?

MATERIALS

- Troll Trap planning sheet (p. 57)
- boxes, plastic containers, oatmeal canisters, juice cartons, foam and paper plates, foil bread pans, large plastic jars, cardboard, strawberry baskets, mesh produce bags, large zippered bags, paper bags
- shredded paper strips, straws, craft sticks, cardboard tubes, plastic knives and spoons, string, yarn, rubber bands, tape
- stuffed animals or dolls

Name: _____

Covered Bridge

Challenge The goats need a safe bridge to cross the river!
Design a covered bridge that will keep the troll away. Then build
and test your bridge.

MY IDEA _____
_____

My Plan (Draw here.) | **Materials I Need**

Did my idea work?

yes

no

How can I make my design better?

Name: _____

A Goat Boat

Challenge Make a goat boat! Design a boat that will carry all three goats across the river safely. Then build and test your boat.

MY IDEA

_____

_____

My Plan (Draw here.)	**Materials I Need**

Did my idea work?

yes

no

How can I make my design better?

Troll Trap

Challenge The three goats wish the troll could be taken far away.
Design a trap that will capture the troll but not hurt him. Then build
and test your trap.

MY IDEA _____

My Plan (Draw here.)

Materials I Need

Did my idea work?

yes

no

How can I make my design better?

Johnny Appleseed

From his first day of life, Johnny Appleseed loved apples. He also loved animals. And animals loved Johnny.

As Johnny grew, he dreamed of planting apple seeds. Johnny always kept the seeds from his apples. He let the seeds dry in the sun. Then he packed them into sacks.

One day, Johnny set out with his seed sacks. As he went, Johnny gave apple seeds to everyone he met. "Put these seeds in the ground," he told them. "Trees full of yummy apples will grow from the seeds."

Johnny also planted his seeds along the way. He planted them in large sunny spaces. "Apples will grow here for anybody who passes by," he said. "And the animals will have apples, too."

At times, Johnny stopped to help a hurt animal.
One day, he helped a wolf get its leg out of a trap.
From that day on, Johnny and the wolf were best friends.

For many years, Johnny walked the land. He gave away hundreds of apple seeds. He planted hundreds of apple trees.

Johnny also made hundreds of friends. People were happy to see him everywhere he went. The animals were happy to see him, too!

Today, Johnny Appleseed is no longer alive. But many say his spirit still lives. It lives in the apple trees that grow all around.

Johnny Appleseed

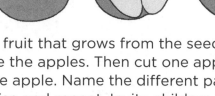

INTRODUCTION

Read aloud the folk tale. Talk about what happens in the story.

To jumpstart scientific thinking, show children three apples with their stems intact. Point out that this is the fruit that grows from the seeds that Johnny Appleseed planted. Have children describe the apples. Then cut one apple in half from top to bottom. Show children the inside of the apple. Name the different parts: skin, flesh, and seeds. Cut another apple in half crosswise and repeat. Invite children to share their comments and observations about apples.

Continue with questions, such as:

- How might Johnny pick apples that he can't reach?
- Why might Johnny want others to plant and grow apples?
- What are some things that we get or can make from apples?
- Is there a way to make apples weigh less so it's easier to carry a bunch of them?

ACTIVITY 1
Apple Picker

Challenge: Design a picker that Johnny can use to gather apples.

Tell children they will make a picker that can be used to pick up plastic apples. Have children write their ideas and draw their plans on their planning sheet. Then have them do the activity. Afterwards, have them complete the rest of their sheet. Discuss which ideas worked, which didn't, and how they can improve their designs.

DIRECTIONS

 Gather the materials you need.

 Build your apple picker. Use your plan from your planning sheet.

 Test your picker. Try to pick up a plastic apple or other plastic fruit with it.

4 Does your picker work? If not, how can you change it to make it work?

MATERIALS

- Apple Picker planning sheet (p. 62)
- sturdy cardboard strips, plastic and foam cups, strawberry baskets, cardboard tubes, wide craft sticks, sturdy straws
- paper fasteners, tape, short lengths of wooden skewers
- plastic apples, oranges, peaches, and other fruits

StoryTime STEM: Folk & Fairy Tales © Scholastic Inc.

ACTIVITY 2

Apple Juice Press

Challenge: Explore how apple juice is made.

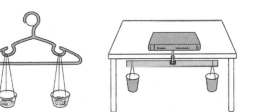

Tell children they will see how much liquid they can mash out of an apple to make apple juice. As children do the experiment, have them fill in their recording sheet. (Note: Tell children they should not eat or drink any part of their experiment.)

DIRECTIONS

1 Put about ½ cup of apple chunks into the plastic bag. Press the air out of the bag. Then seal the bag tightly.

2 Mash the apple chunks as flat as possible. Use the rolling pin or wooden block. What happens? Record on your recording sheet.

3 Place the strainer on an empty tumbler. Empty the bag into the strainer. Use a spoon to mash the apples a little more. What happens? Record on your recording sheet.

4 Pour the apple juice into a measuring cup. How much juice did you make? Record on your recording sheet.

MATERIALS

- Apple Juice Press recording sheet (p. 63)
- thick, quart-size, zippered plastic bags
- peeled apples, cut into small chunks
- rolling pins, wooden cylinders and blocks
- clear plastic tumblers
- sturdy plastic spoons
- strainers, measuring cups

ACTIVITY 3

A Lighter Apple

Challenge: Make a balance scale to weigh apples.

Apples can get very heavy. Tell children they will find out how to make an apple weigh less. But first, they will build a balance scale to weigh an apple. Have children write their ideas and draw their plans on their planning sheet. As children do the experiment, have them fill in their sheet. Afterwards, discuss with children what they think happened to the apple slices.

DIRECTIONS

1 Gather the materials you need.

2 Build your balance scale. Use your plan from your planning sheet.

3 Put an apple slice on one side of your scale. Add objects on the other side. (The objects should all be the same.) Keep adding until both sides are even. How many objects does the slice weigh? Record on your sheet.

4 Pull a string through the slice. Hang it in an open space. Wait one week.

5 Weigh the slice again. Use the same objects you used before. How many objects does it weigh? Record on your sheet.

MATERIALS

- A Lighter Apple planning sheet (p. 64)
- **balance scale:** rulers, yardsticks, wooden dowels, clothes hangers; string, twine, yarn; empty margarine tubs, plastic cups, chip canisters; tape, hole punch
- **objects to weigh:** crayons, paper clips, pencils, cubes
- peeled and cored apples, sliced into rings
- string, nails or hooks (for hanging the apple slices)

Apple Picker

Challenge Johnny picks apples all day. Design an apple picker to help make his task easier. Then build your picker and test it.

MY IDEA

_____

_____

My Plan (Draw here.)	**Materials I Need**

Did my idea work?

☺ yes

☹ no

How can I make my design better?

Apple Juice Press

Challenge Make apple juice! How much juice can you get from an apple?
Find out here.

Do the activity. Fill out each section below to show what happens.

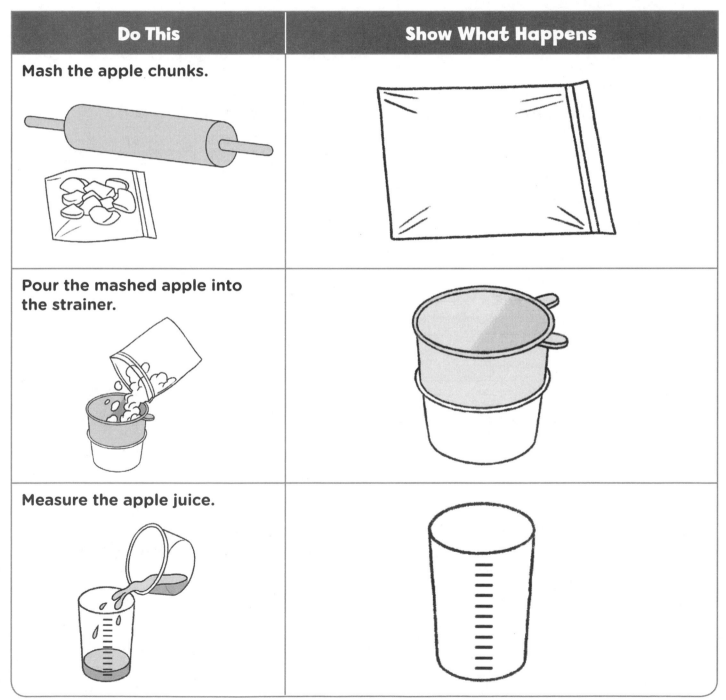

Do This	Show What Happens
Mash the apple chunks.	
Pour the mashed apple into the strainer.	
Measure the apple juice.	

A Lighter Apple

Challenge Design a balance scale to weigh an apple.
Then find out how to make an apple lighter.

MY IDEA _____
_____

My Plan (Draw here.)	**Materials I Need**

Objects I used to weigh the apple slice:

How much does the apple slice weigh?	Wait one week. How much does the apple slice weigh now?

StoryTime STEM: Folk & Fairy Tales © Scholastic Inc.

Paul Bunyan

Paul Bunyan was born a BIG baby!
He slept in a big crib.
He rode in a big stroller.
He played in a big playpen.

Paul grew and grew.
He grew too big for his house.
So Paul had to move out into the woods.

One day, Paul pulled a big, baby ox out of the snow.
He named the ox Babe.

Paul took good care of Babe.
The ox grew and grew.
Soon Babe was a very BIG ox!

Paul and Babe became a team.
They did BIG things together.
They cut down big trees.
They made big fields for the farmers.

Paul and Babe had a big logging camp. Every day, Paul cooked hotcakes for his workers. He used a pan as big as a football field. Men on bikes took the hotcakes to the workers. They rode their bikes up and down a big, long table.

Once, Paul and Babe dug five big holes. They filled the holes with water. We call these the Great Lakes today.

Another time, Paul and Babe walked across the West. Paul dragged his ax part of the way. He left a big ditch behind him. We call that ditch the Grand Canyon today.

Paul and Babe did many big things in their life.
So where did they go to live in their old age?
The BIG state of Alaska, of course!

Paul Bunyan

INTRODUCTION

Read aloud the folk tale. Talk about what happens in the story.

To jumpstart scientific thinking, display the art that comes with the story. Point out Paul Bunyan's large size. Ask children what problems might Paul Bunyan's family have faced as he kept growing. Invite them to share possible solutions to some of those problems. Then talk about why Paul moved out into the woods and how this may have made his life easier or harder.

Continue with questions, such as:

✿ In what ways might Paul's size be helpful to him and others? How might it be harmful?

✿ How might Paul get pancakes to his workers without using bikes to deliver them?

✿ What are some other ways Paul and Babe might have changed the land or the environment?

ACTIVITY 1
Swinging Hammock

Challenge: Make a hammock for young Paul Bunyan to sleep in.

Tell children they will build a hammock for young Paul Bunyan. The hammock should be able to support a stuffed animal or doll, even when it swings back and forth. Have children write their ideas and draw their plans on their planning sheet. Then have them do the activity and complete their sheet. Discuss which ideas worked, which didn't, and how they can improve their designs.

DIRECTIONS

1 Gather the materials you need.

2 Build your hammock. Use your plan from your planning sheet.

3 Choose a stuffed animal or doll to represent Paul Bunyan.

4 Put Paul in the hammock. Does he fit in it? Can it support his weight?

5 Now, swing the hammock back and forth. Does Paul stay in the hammock? If not, how can you change it to make it work?

MATERIALS

- Swinging Hammock planning sheet (p. 69)
- **hammock:** old tablecloths, sheets, towels, shirts and pants, shower curtains, mesh produce bags, sheets of crepe paper, plastic shower-curtain rings, plastic six-pack soda rings, metal canning jar rings
- **support:** broomstick, large dowels, chairs, tables, easel stands, short bungee cords, thin rope, cording, fabric ribbon
- duct tape, stapler, hole punch, pipe cleaners
- large stuffed animals or dolls

StoryTime STEM: Folk & Fairy Tales © Scholastic Inc.

ACTIVITY 2
Hotcake Catapult

Challenge: Design a catapult that Paul can use to serve hotcakes to workers at the other end of the table.

Tell children they will build a hotcake catapult that Paul can use to fling hotcakes to his workers sitting around a large table. Have children write their ideas and draw their plans on their planning sheet. Then have them do the activity and complete their sheet. Discuss which ideas worked, which didn't, and how they can improve their designs.

DIRECTIONS

 Gather the materials you need.

 Build your catapult. Use your plan from your planning sheet.

3 Cut out a cardboard circle to represent a hotcake. Place it on your catapult.

4 Now, use your catapult to send the hotcake down the table. Does your catapult work? If not, how can you change it to make it work?

MATERIALS

- Hotcake Catapult planning sheet (p. 70)
- craft sticks, sturdy cardboard, plastic tubes, empty soda bottles of various sizes, shallow plastic dishes, old plastic gift cards, strips of thick cardboard, plastic spoons, knives, and forks, wooden clothespins, rulers
- rubber bands in various sizes
- cardboard, scissors, tape

ACTIVITY 3
Build a Tower

Challenge: Build a tower that Paul's friends and family can use when they want to talk to him face to face.

Tell children they will build a tower for Paul's friends and family so they can talk to him face to face. Explain that the tower should be sturdy enough to hold at least one doll or toy that represents a friend. Have children write their ideas and draw their plans on their planning sheet. Then have them do the activity. Afterwards, have them complete the rest of their sheet. Discuss which ideas worked, which didn't, and how they can improve their designs.

DIRECTIONS

 Gather the materials you need.

2 Build your tower. Use your plan from your planning sheet.

3 Choose a small doll to represent one of his friends. Place the friend in the tower.

 Does the tower support his friend? Does it stand firm? How many more friends can the tower hold?

MATERIALS

- Build a Tower planning sheet (p. 71)
- tall cardboard boxes (cracker or cereal boxes), sturdy cardboard, tall chip canisters, paper towel tubes, large plastic cups, long sturdy straws, long sturdy cardboard strips, craft sticks
- tape, scissors
- toys or dolls

Swinging Hammock

Challenge Young Paul Bunyan needs a safe place to sleep. Design a hammock that will hold him, even when it swings back and forth. Then build and test it.

MY IDEA

_____

_____

My Plan (Draw here.)

Materials I Need

Did my idea work?

yes

no

How can I make my design better?

Name: _____

Hotcake Catapult

Challenge Paul needs a way to serve hotcakes to workers at the other end of the table. Design a hotcake catapult for Paul. Then build your catapult and test it.

MY IDEA

_____

_____

My Plan (Draw here.)	**Materials I Need**

Did my idea work?

yes

no

How can I make my design better?

Name: _____

Build a Tower

Challenge Paul's friends and family want to talk to him face to face. But Paul is very tall! Design and build a tower that can hold at least one friend.

MY IDEA

_____

_____

My Plan (Draw here.)

Materials I Need

Did my idea work?

yes

no

How can I make my design better?

John Henry

John Henry was born a strong baby. He was also born with a hammer in his hand.

John Henry loved to hammer. He dreamed of using his hammer to build railroads.

One day, John Henry met the railroad boss. He showed the boss his strong hammer swing. The boss gave John Henry a job right away.

John Henry's job was to hammer and drill.
He hammered spikes into the side of a mountain.
Then he turned the spikes to drill holes in the rock.
John Henry was the strongest, fastest man on the job.

StoryTime STEM: Folk & Fairy Tales © Scholastic Inc.

One day, a city man came with a drilling machine. He said his machine could beat even the strongest, fastest man. So the railroad boss set up a contest. It would be between John Henry and the machine.

People came from all around to watch the contest. "Ready. Set. Go!" yelled the railroad boss. And the contest began.

John Henry hammered and drilled. The drilling machine rumbled and buzzed.

At times, the machine was winning. Then John Henry was winning. And so it went, all day long.

Just before the sun set, the machine broke down. The contest was over. John Henry was the winner!

John Henry held up his hammer. He smiled his biggest smile. The crowds cheered.

Just then, John Henry fell to the ground. His strong heart had stopped beating.

On that day, John Henry's life ended just like it began—with a hammer in his hand.

John Henry

INTRODUCTION

Read aloud the folk tale. Talk about what happens in the story.

To jumpstart scientific thinking, display the "Simple Machines" learning poster (p. 79). Review each of the simple machines and point out the examples shown on the poster. Invite children to name other simple machines and talk about how they work. Then invite children to share their thoughts about different kinds of simple machines that John Henry might use in his work.

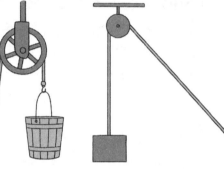

Continue with questions, such as:

- ✿ How might John Henry move away heavy rocks that block his way?
- ✿ How might a crane be helpful to John Henry?
- ✿ What does your heart do? Why is it an important part of your body?
- ✿ Why do you think John Henry's heart stopped?

ACTIVITY 1
Rock-Lifting Crane

Challenge: Design and build a crane that lifts, lowers, and moves rocks.

Explain to children that as John Henry drilled through the mountain, a lot of rocks needed to be moved. Tell children they will build a crane that can lift and lower a pretend rock and move the rock to one side. Have children write their ideas and draw their plans on their planning sheet. Then have them do the activity. Afterwards, have them complete the rest of their sheet. Discuss which ideas worked, which didn't, and how they can improve their designs.

DIRECTIONS

1 Gather the materials you need.

2 Build your crane. Use your plan from your planning sheet.

3 Ball up a piece of foil to use as a rock. Place the rock in the crane claw.

4 Now, test your crane. Does it lift and lower the rock? Does it move the rock to one side? If not, how can you change it to make it work?

MATERIALS

- Rock-Lifting Crane planning sheet (p. 76)
- **crane's base:** tall chip and oatmeal canisters, juice cartons, shoeboxes, plastic soda bottles, stacks of blocks and books
- **crane:** craft sticks, sturdy cardboard strips, sturdy straws, wooden dowels, thread and ribbon spools, thin rope, ribbon, cording, pipe cleaners
- **rocks:** recycled paper and foil (balled up)
- paper fasteners, tape, scissors

ACTIVITY 2
Rock Dumpster

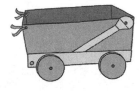

Challenge: Create a dumpster that can be raised to dump out rocks.

Tell children they will build a rock dumpster. Explain that the dumpster should hold rocks and then lift up so the rocks slide out when it is being emptied. Have children write their ideas and draw their plans on their planning sheet. Then have them do the activity. Afterwards, have them complete the rest of their sheet. Discuss which ideas worked, which didn't, and how they can improve their designs.

DIRECTIONS

1. Gather the materials you need.

2. Build your rock dumpster. Use your plan from your planning sheet.

3. Make paper or foil "rocks." Put the rocks in your dumpster.

4. Now, raise your dumpster to empty out the rocks. Does it work? If not, how can you change it to make it work?

MATERIALS

- Rock Dumpster planning sheet (p. 77)
- **dumpster box:** cardboard boxes, shoeboxes with lids, juice cartons, empty food boxes, cardboard trays
- **wheels and axles:** chip canister lids, thick cardboard circles, thread and ribbon spools, wooden dowels, straws
- craft sticks, thick cardboard strips
- **rocks:** recycled paper and foil (balled up)
- paper fasteners, tape, scissors

ACTIVITY 3
Hammering Heart Rate

Challenge: Measure heart rate and see how it changes.

Tell children that they can find out how fast their heart beats by measuring their heart rate. Show them how to check their heart rate using the pulse at their wrist. As children do the activity, have them fill in their recording sheet.

DIRECTIONS

1. Do each activity listed in your recording sheet.

2. Find your pulse on your wrist. Feel for a small beat under your skin.

3. Count how many beats you feel in 1 minute. Ask a partner to time you, using a stopwatch.

4. Fill in your heart rate.

5. What do you notice about your heart rate?

MATERIALS

- Hammering Heart Rate recording sheet (p. 78)
- a partner
- stopwatch

Name: _____

Rock-Lifting Crane

Challenge John Henry needs to move some rocks. Design a crane that can lift, lower, and move rocks. Then build and test it.

MY IDEA _____
_____

My Plan (Draw here.) | **Materials I Need**

Did my idea work?

yes

no

How can I make my design better?

Name: _____

Rock Dumpster

Challenge Help John Henry dump rocks away from his work area. Design a rock dumpster that can be raised and lowered. Then build and test your dumpster.

MY IDEA _____
_____

My Plan (Draw here.)

Materials I Need

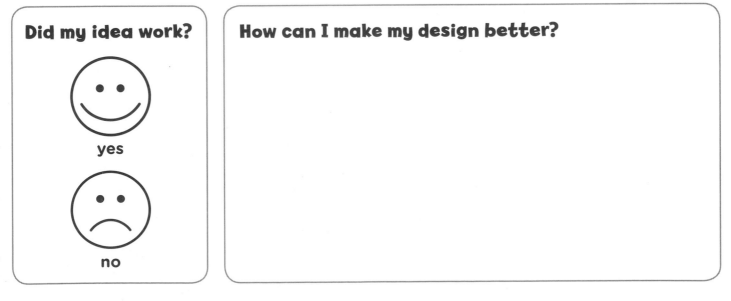

Did my idea work?

yes

no

How can I make my design better?

Name: _____

Hammering Heart Rate

Challenge Measure your heart rate when you are resting. Then exercise and measure it again. What do you notice?

Activity	Heart Rate	My Observations
Rest: Sit quietly for 1 minute.	_____ beats per minute	
Exercise: Be John Henry! Swing a pretend hammer for 1 minute.	_____ beats per minute	
Rest: Sit quietly again for 1 minute.	_____ beats per minute	

Simple Machines

**Simple machines help make work easier.
They have only a few or no moving parts.**

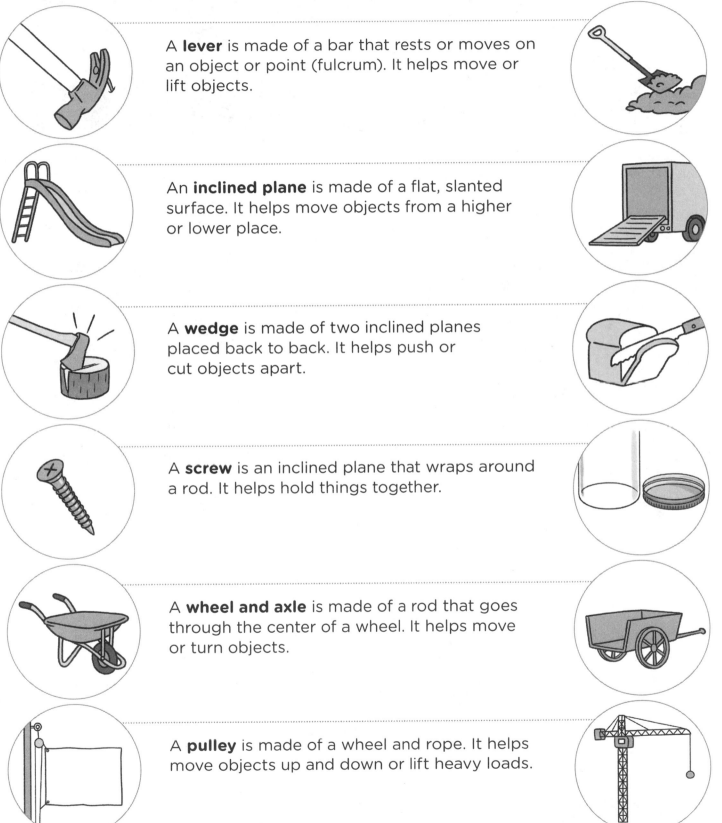

A **lever** is made of a bar that rests or moves on an object or point (fulcrum). It helps move or lift objects.

An **inclined plane** is made of a flat, slanted surface. It helps move objects from a higher or lower place.

A **wedge** is made of two inclined planes placed back to back. It helps push or cut objects apart.

A **screw** is an inclined plane that wraps around a rod. It helps hold things together.

A **wheel and axle** is made of a rod that goes through the center of a wheel. It helps move or turn objects.

A **pulley** is made of a wheel and rope. It helps move objects up and down or lift heavy loads.

My Folk & Fairy Tales
STEM Journal

by _____